Hyrule Warriors: Age of Calamity

Guide - Walkthrough, Tips, Tricks, And More!

By Ibram X.L

Things the Game Doesn't Tell You

Try not to Fret Over Understanding the Blacksmith at First

The Hylian Blacksmith Guild is the principle strategy for improving the intensity of your champions as you progress through the game. Melding weapon drops you get while finishing missions at the Blacksmith lets you engage different weapons by either giving them experience, making them hit more enthusiastically, or assimilate new powers.

The instructional exercise about how it functions is tossed at you from the get-go in the game while you're actually getting a hang of things, and it is highly unlikely to return to it later. It's anything but difficult to feel overpowered. Utilize Link's beginning weapon to rehearse how the Blacksmith functions. Try not to stress over creation a misstep you can generally fix it later. You'll be supplanting that beginning weapon with something that is associations better in Chapter 3 at any rate.

When you finish Chapter 2 you'll have a decent handle on how Hyrule Warriors functions. This will assist you with bettering comprehend the weapon customization in more important manners.

In the wake of finishing the primary Chapter 5 mission, "???", you will have the option to finish the journey "Thinking about the Statues" in Gerudo that will permit you to eliminate one seal that you've joined to weapons all at once. You gain the capacity to eliminate all seals on a weapon in the wake of completing "Mystery of the Rito Artisans" in Hebra.

Giving Orders

Most missions with an enormous front line have various saints follow along to the battle. The mission will at that point have numerous destinations on the guide to finish simultaneously. You can hurry to every target yourself with your #1 saint, however that can destroy important time on the clock. All things being equal, issue requests to your different fighters on the delay screen.

On the respite screen you'll get an enormous guide of the front line. From here you can give requests to different heroes, sending them to an area you pick. While you're battling ceaselessly in one zone, that fighter will move to where you sent them, overlooking foes en route. At the point when they show up they'll battle anything close by, however they do negligible harm until you assume responsibility for the champion. Whenever you've completed the goal you're battling at, press Up or Down on the D-Pad to change to the next legend who is as of now at the following spot. This spares you time by putting different champions where you need them next while you battle away.

Doing this decreases the number of materials you get for making and missions later. It's dependent upon you to find that balance, however on simpler challenges the game is fairly lenient and you can change things up with no punishment in the respite menus.

You Can Ignore Capturing Outposts Unless Required

The enormous combat zone maps regularly have various stations that you can catch away from the foe crowds. While it feels great on a completionist level, and you get extra creating and journey segments, except if the mission objective explicitly states you have to keep or hold a base, you can skip it completely.

Wands Are Extremely Useful

Wizzrobes litter the war zones of Hyrule and they regularly

feel like a disturbance. They're not excessively troublesome, however they are either in the path or there to gobble up the check in missions where time is short. Their mystery however is that they are around to help give you an extraordinarily significant apparatus against supervisors, wand energy. There are three kinds of Wizzrobes: fire, ice, and lightning. Murdering one drops things that reestablish your wand energy for that component.

Utilizing a wand on non-natural foes stops them for a brief timeframe, otherwise called amazing them, permitting you to work on their obstruction for a second. In case you're ready to stop an essential adversary with its contrary component, for instance, utilizing the ice wand on a foe based ablaze, you'll debilitate their hindrance altogether, permitting you to tear through it, at times quicker than setting off a Flurry Rush.

You can adequately stop extreme foes like the Lynel (centaur) to an end with a wand with staggered employments of a wand and they won't have the option to do anything.

Take as much time as is needed on Boss Fights

Besides during side journeys, most supervisor fights, what I'm calling any battle with a named animal, are not planned. This is the ideal occasion to figure out how they battle, to get familiar with their signs for which Shiekah Slate capacities to utilize and when, and to rehearse evade timing to trigger Flurry Rush.

Similar managers come up a great deal as you progress through the game, either in the story or in side missions, so it will be exceptionally gainful to realize what certain adversaries do from the get-go.

Connection utilizing his own wellbeing to engage Strong Attacks when utilizing a two-gave weapon.

Give Link a Different Weapon

Connection appears to be an exceptionally essential legend from the start with his customary blade and shield, yet he has a great deal of variety since you can give him a plenty of various weapons.

Two-gave weapon: Link can supplant his Strong Attacks with his new Unique Ability that rather utilizes his own wellbeing with much more remarkable Strong Attacks. The hearts you use to fuel this capacity will turn white, and in the event that you can stop and pause for a minute to squeeze X to eat food before you get hit, you'll get those hearts back. It's a bet in light of the fact that the more you abandon eating, the more grounded these assaults are.

Lances: Spears permit Link to make since quite a while ago ran skirmish assaults and to charge through swarms of adversaries, both with his Strong Attacks and his Unique Ability.

Boomerang: This works almost indistinguishable from your blade and shield. As a matter of fact, at one point I had the option to get Link to toss it and it returned, yet I have not had the option to reproduce this. Any assistance from the network in the remarks would be valued.

Request Allies Around

Toward the beginning of a fight, if there is a partner around, consistently make a point to arrange them the other way or zone that you are planning to go. Having at least one partners dissipated over the guide will make things go a lot quicker.

This is on the grounds that players can twist between characters on the fly, making it far easier to cover enormous guides. This even works for community, despite the fact that at any rate three legends should be on a guide so it doesn't generally work in Hyrule Warriors: Age Of Calamity.

Spare Meals For Big Battles

It very well may be enticing to eat before each mission to support protection or experience gain, however food supplies will go quick. That is the situation for the early hours at any rate.

The best tip is to spare dinners for story missions, as those will in general be the most burdening. The side fights are regularly easier or short enough where these food supports aren't required.

Get Link Into His Boxers

This may seem like a bizarre tip yet it will make Hyrule Warriors: Age Of Calamity endlessly more clever. Seeing Link snort in light of his partners while in his fighters is something to see.

Hyrule Warriors: Age Of Calamity needs somewhat more craziness so why not add to it? Apparel, all things considered, doesn't generally help guard as it does in Breath of the Wild, in this way losing garments won't be unfavorable to fight.

Try not to Neglect Tools

The apparatuses, similar to the bombs, were extraordinary for puzzle understanding in Breath of the Wild yet they weren't generally hostile weapons. The inverse is valid in Hyrule Warriors: Age Of Calamity since they are presently more valuable than any other time in recent memory.

In addition, their material science are as yet flawless; for instance, utilizing ice on water will sling Link into the air. It should likewise be noticed that each character utilizes these instruments in an unexpected way, making for remarkable combos.

Search For Hidden Objects

Missions are not only there to wreck tons of foe troops. There are likewise insider facts to reveal on each guide. There are money boxes that are covered which can be unearthed through the magnet device.

There are additionally Korok Seeds, in spite of the fact that their numbers are not the same number of as Breath of the Wild. While you are pummeling enemies, be watching out for the phenomenal.

How Co-Op Works

Community can be turned on or off whenever in Hyrule Warriors: Age of Calamity and for any mission inasmuch as it has two characters. The instructional exercise stage from the beginning cannot be played in community, for instance.

Center, for the present, is nearby just, which is useful for families cooped up in isolate however not very good for gamers that need to connect with companions. Will Nintendo add online help? That stays a secret for the present.

Amiibo Support

Hyrule Warriors: Age Of Calamity apparently bolsters all Amiibo Nintendo has put out hitherto. A Zelda related figure should dole out better rewards however.

In contrast to Breath of the Wild, where the measure of Amiibo somebody could use in a solitary day was limitless, Hyrule Warriors: Age Of Calamity confines use to five figures. Regardless of whether you're not anticipating playing, you ought to at any rate examine an Amiibo to develop those lifts.

Try not to Waste Rupees On Weapons

There is a combination framework in Hyrule Warriors: Age Of Calamity that permits players to fundamentally step up weapons utilizing others as impetuses. This can get costly and just gives a negligible prize.

Rupees are better spent in shops getting elements for cooking or for quests. Opening journeys to construct wellbeing or combo meters up is unquestionably more valuable than the weapon combination.

Remember To Glide

Floating can be somewhat dubious to pull off. It's somewhat simpler for Link since he can utilize that ice column to give himself a lift. Different characters can attempt to bounce off dividers and afterward skim in midair.

Height assumes a major part in progress rates. In the event that there is an incline around, consistently ensure and skim away for snappy vehicle. Careful discipline brings about promising results, and it may take several attempts to get the hang of gliding.

Foes Are Currency, Don't Ignore

It very well may be enticing to go directly to destinations in each mission, yet killing foes is likewise helpful for a few reasons. Initially, they are worth experience focuses and they drop things. Furthermore, they are additionally money.

It's just plain obvious, in the event that you kill a Moblin, at that point you get a Moblin abundance card. These are utilized to open redesigns, for example, for wellbeing. This adds a smidgen more significance to killing the unlimited hordes.

Tips, Tricks and Strategies for New Players

Try not to Button Mash!

The disorder and scene joined with the generally basic battle controls make button pounding enticing. In any case, the way to playing Warriors games viably is playing them productively. Each character has various combos that function admirably in various conditions, and realizing which is best in every situation is fundamental. While these games do zero in on savage power and redundant battle, there is a layer of care that thoughtless catch crushing overlooks. Not exclusively will ponder play help amidst a front line overwhelm with adversaries, it'll make the interactivity all the more captivating as well.

Continually Give Orders and Swap Characters

Notwithstanding learning combos, another disregarded aspect of Warriors titles is the vital component. Most fights give the player control of numerous commanders that can be traded whenever. This isn't only for changing it up however, as performing various tasks in Age of Calamity is crucial. Regularly, numerous focuses on the guide will be assaulted without a moment's delay, and it's dependent upon the player to shuffle every one of these purposes of contention. By utilizing the guide, you can appoint the chiefs to various assignments and cycle between them to cover the war zone.

The game pushes the major part toward this path; however, it doesn't pressure that it is so critical to consistently be turning among warriors and providing orders. In certain regards, is comparable to Pikmin 3 Deluxe, where time and asset the board are fundamental. By keeping your group moving to various focuses over the field, you can keep the adversary from catching fortifications and clearing out crews of cordial AI all the more effectively. You ought to consistently have a large portion of your consideration on the guide and where chiefs are and flipped between them varying.

Utilize Rods and Runes

In the first Hyrule Warriors, the player approached conventional Zelda things, for example, the Hookshot and Boomerang. In Age of Calamity, the player approaches the Sheikah Runes from Breath of the Wild, just as different natural bars. While it tends to be barely noticeable these, they can truly switch things around. Numerous adversaries have shortcomings to one component or thing, and utilizing them can uncover their powerless point checks. Once more, productive battle is focal, so trying different things with these devices to sort out their uses is beneficial.

Pick Characters with Engaging Playstyles

Like Super Smash Bros., all the characters in Age of Calamity utilize a similar control arrangement and essential combos. Nonetheless, they all have particular styles, with extraordinary capacities planned to ZR. These regularly educate different moves and how they chain together. Ordinarily, the player can pick which characters to bring into fight, so it merits trying different things with each to figure out which playstyle feels the best. There truly is no off-base answer regarding which characters to bring into a specific battle, so it's generally an issue of player inclination.

Be Patient When Fighting Bigger Enemies

While the more modest adversaries on the field can be cut down effectively, adversary commanders and other additionally forcing enemies can't be. Basically, utilizing great combos and things won't sufficiently be, as they'll do moderately limited quantities of harm. The situation here is tolerance, ingenuity and example acknowledgment.

It's crucial that the player opens up bigger adversaries by striking at their frail focuses, which just become accessible by learning assault examples and hanging tight for the correct chance. In these cases, it's imperative to stand by and assault when the powerless focuses are uncovered, as doing whatever else can prompt superfluous harm and knockback that could drive you to miss an opening.

Exploit the Hyrule Map

The center of the game is the fights themselves, yet between experiences, the player can spend assets to satisfy assignments on the overworld. These are fairly latent approaches to draw in with Hyrule, but they are totally worth doing. Prizes range from new combo trees to sellers that can overhaul gear. This plunder can be fundamental to amplifying involved with where harder fights aren't so overwhelming. Furthermore, huge numbers of the assets expected to finish these missions are inactively gotten in fight, so there's no reason not to finish these undertakings.

GameSpot may get a commission from retail offers.

Hyrule Warriors: Age of Calamity is at long last out on Nintendo Switch and has a lot harder fights than what you may have looked in the free eShop demo. To assist you with a portion of the difficulties that lie ahead, we've spread out some helpful battle tips to raise your aptitudes.

We've additionally discovered some energizing tech that will be extra valuable for the individuals who need to get profound into the complexities of battle. So, this will likewise be a guide containing some serious tips from our inhabitant Breath of the Wild master Max Blumenthal, who has a skill for finding new battle tech, and has just spearheaded a couple of Age of Calamity stunts!

On the off chance that you have additional tips, post them in the remarks underneath. We're continually searching for cool battle strategies to use in Hyrule Warriors, so don't spare a moment to get down on them. Something else, read our full Hyrule Warriors: Age of Calamity audit.

Use Stasis Correctly

In case you're tossing out Stasis harum scarum and utilizing its auto-movement, you must alter your way of life. While utilizing Stasis to counter a Stasis-feeble assault is a decent and common use of the capacity, the most remarkable portion of Stasis is its capability to broaden powerless point term and stagger foes for more. You can exploit this by running subsequent to performing Stasis, permitting you to then uninhibitedly do another activity. While some powerless focuses last some time for you to pummel an adversary, others keep going for one moment or two, which isn't sufficient opportunity to break it. So use Stasis to expand that time.

Hyrule Warriors: Age Of Calamity 6 Advanced Combat Tips And Tricks

This strategy is vital to broadening combos, increasing more supermove meter, and confining a foe's capacity to retaliate. Sadly, few out of every odd character can assault during Stasis- - Daruk, for instance, can't catch up with an assault. In any case, for characters who can, this procedure is amazingly incredible. It can even be hung into adversary powerless focuses you trigger with Sheikah Slate capacities, permitting you to drive your way through a rival's frail point whenever.

When in doubt, you'll need to hit foes subsequent to setting off this method however much as could reasonably be expected to amplify your harm yield, working towards that unavoidable feeble point crush. It proves to be useful, particularly against more impressive foes who will in general be very tanky in harder troubles. So, don't belittle this procedure when there's no other option!

How to Best Use Elemental Rods

Basic Rods are basic to dominating Age of Calamity's battle. These incredible mystical weapons effectively make feeble focuses whenever in case you're experiencing difficulty making an initial yourself, yet the sort of frail point and size of the impact relies upon a few things.

For one, it's imperative to take note of that assaulting basic based adversaries with wands of a similar component do no harm, so don't hit an ice foe with an Ice Rod.

The force and size of the Elemental Rod impact will rely upon the landscape you hit.

Second, hitting adversaries utilizing a component it's powerless against (fire against ice, ice against fire, lightning against foes absorbed water) consequently triggers a basic feeble point. This incredible feeble point assault makes a more extended enduring and more fragileweak point measure.

Ultimately, the force and size of the Elemental Rod impact will rely upon the territory you hit. You can see this while drifting your fire bar over grass rather than earth. Utilizing fire on green regions, lightning on metal or water, and ice on water will expand the impact reach and power a basic feeble point on foes trapped in the impact, even to those not conventionally frail to them. So make certain to consistently focus on your environmental factors and exploit the territory you're on!

Depend On Magnesis Cancel For A Quick Dodge Or Attack Cancel

Most characters can't utilize Magnesis if nothing metallic is near. Also, if there are metallic things around, you'll play out a normal Magnesis assault, which is pretty disappointing for most characters. Anyway, you may be reasoning: "Admirably, that is feeble, so what's the point?"

Yet, in reality, Magnesis is seemingly one of the more remarkable strategies in the game since you can utilize it to drop practically any activity you do. It likewise gives you complete power without utilizing assets.

Magnesis is apparently one of the more impressive methods in the game since you can utilize it to drop practically any activity you do.

You can play out a Magnesis Cancel by squeezing the R and B catches when there are no metallic things inside reach. On the off chance that a standard evade isn't sufficiently brisk to move you, you can utilize Magnesis Cancel to all the more rapidly avoid assaults or offset longer assault activitys.

You'll need to remember that the adequacy of Magnesis Cancel changes from character to character, as a few, as Zelda, can't do this method by any means. Interestingly, others like Impa can spam it rapidly again and again. Be that as it may, when the circumstance is correct, Magnesis Cancel can be very grip. A few assaults, for example, a Guardian's lasers, have an awesome time hitbox that is difficult to evade and can't be parried with different characters, so keep this in your collection when facing destructive enemies!

Impa is Broken (So Use Her!)

Generally, we're not especially anxious to advance simple successes, but rather it's difficult to overlook when a character like Impa is 100% SSS level on the level rundown. She can perform close unlimited combos on single adversaries and groups while shocking everybody around her. Her meter-gain is additionally staggeringly quick when battling against gatherings, which permits you to play out your exceptional frequently.

In case you're struggling, you can incline toward Impa to help you through a harsh mission.

Later on, it's conceivable that different characters may have shrouded potential that will put them as high on the level rundown as Impa, yet she's right now pretty prevailing. Furthermore, with capacities opened in the endgame, she turns out to be much a greater amount of an outright beast. In case you're struggling with this game, you can incline toward this Sheikah torment train to bring you through an unpleasant mission.

Menu Slowdown Can Help Practice Parrying or Dodging

This method can help when absolutely necessary in case you're making some unpleasant memories repelling or avoiding. By exchanging the Sheikah Slate and Elemental Rod menus, you can place yourself in a ceaseless condition of log jam, giving you a significantly more precise window to evade and repel assaults. Some foe assaults have abnormal movements that are trying to time a repel or avoid against, so on the off chance that you need one moment to look for the counter-assault advise, you can utilize this little stunt to help nail the circumstance.

Other Assorted Tips

As instituted by Max, the Urbosa Float Cancel is a mystery battle stunt that can be executed while airborne and assaulting. By tapping your ZR charge, your assault check will be reset, permitting you to keep assaulting noticeable all around. This flying Urbosa stunt will keep on headshot most foes while being out of their reach and as yet doing harm. It's conceivable to get hit by high assaults, yet this strategy regularly evades most ground assaults while ceaselessly doling out harm from above.

By doing your C4 combo (Y-button x4 + X-button) noticeable all around and timing a Sheikah Slate capacity consummately, you can slide along the ground dangerously fast.

Max has additionally discovered a remarkable strategy for Revali, which he's named Rune Sliding. By doing your C4 combo (Y-button x4 + X-button) noticeable all around and timing a Sheikah Slate capacity impeccably, you can slide along the ground dangerously fast. Contingent upon your rune of decision, a few slides last more than others, however whichever one you pick, the utility changes with your quick development, and during the length of the movement, you are totally powerful. This strategy can be helpful for moving toward bothersome ran foes or fleeing from a battle.

Planning tips and deceives

Cook the best nourishment for the battle: You'll open more food sources as you complete missions and give materials to various symbols on the guide. Peruse the portrayal for every mission and afterward pick the dishes that will give you the best lifts for those battles.

Work to open the entirety of the characters immediately: Most of the playable characters open by essentially working through the fundamental missions, yet you'll have to utilize the Military Training Camp to make them more grounded.

Utilize the Military Training Camp to step up: By spending rupees, you can rapidly step up individuals from your gathering at the Military Training Camp arranged in the upper piece of Central Hyrule. You'll simply have the option to level characters up as high as your most elevated level character.

Breaker your weapons at the Hylian Blacksmith: You can discover the Hylian Blacksmith directly close to the Military Training Camp on the guide. This element permits you to consolidate the intensity of various weapons together to make one more grounded weapon. Indeed, this costs rupees, however it will make it simpler for you to crush your adversaries.

Combat tips and tricks

Leave the mission on the off chance that you realize you won't make as far as possible: There's no motivation to exhaust your thumbs for reasons unknown. In the event that you can tell that it is highly unlikely your characters can meet a period limit with their present level and weapons at that point press the - catch and afterward select Exit Scenario to make a beeline for the guide before the check runs out.

Top off your capacity guage by crushing more modest adversaries: Whenever a character's capacity guage tops off, you can deliver a devestating assault by squeezing the A catch. To top it off rapidly, rout minor adversaries in the region.

Bring down a named adversary's Weak-point guage: More impressive foes will have a Weak-point guage that gradually brings down as you assault. In the event that you can totally separate it they will be paralyzed incidentally. Utilize the time they are shocked to deliver an amazing A catch assault.

Request your characters to various areas on the guide: While undertaking a mission, you can order your characters to various areas on the guide. That way, you can rapidly bounce among them and annihilation adversaries at various destinations faster.

Exploit Sheikah Slate Rune shortcomings: Occaisionally while you battle an enormous foe will develop to do an assault and a Rune image will show up over its head. Promptly pull out your Sheikah Slate and assault with the relating Rune to briefly shock it.

Evade as opposed to utilizing your Shield: You'll actually try not to take harm and you may get remunerated with a Flurry Attack opportunity where you can land a lot of hits on the adversary without them having the option to hit back.

Item and material tips

Try not to get things: Whenever a Rupee, thing, or collectable drops from a Chest or adversary it will naturally get added to your stock after a limited quantity of time, so don't sit around going around and getting them yourself.

Sort out the area of explicit materials with the Sheikah Sensor: You open the Sheikah Sensor in the wake of finishing the mission named "Required: Researchers!" Once opened, at whatever point you drift over a mission that you're feeling the loss of the necessary materials for, basically press the X catch and afterward the areas that will give you the materials you need will get a green hover around them.

Finishing missions opens more redesign alternatives and administrations: Sometimes when you complete missions, regardless of whether large or little, you'll be remunerated with more administrations, stores, or followup missions. Therefore, it's a smart thought to finish the entirety of the missions in all aspects of the guide.

Zelda amiibo give you things: If you have any of the Zelda amiibo you can filter five of them in every day to open free things. It may very well give you what you have to proceed to your next mission.

Switch between characters!

This may appear glaringly evident; however, utilize the characters you've opened. Every warrior has novel moves and an alternate vibe — you can analyze utilizing every one of them in the Meditative Training 'dojo' which opens in the initial phases of the game (hit 'L' to jump to the Services tab and it's the third choice on the rundown), and we suggest heading there with each character for a moment or two to explore and get settled with them. Careful discipline brings about promising results.

Past that, it's frequently substantially more helpful (also time productive and fundamental at higher troubles) to switch characters who are over the guide as opposed to sit around idly gallivanting your significant level Link over the whole war zone while the remainder of your group gets destroyed. Best take a note from your number one Nintendo comfort and get switchin'.

Pay to step up characters and try not to pound

The contender you're utilizing will accumulate XP through fights and level up normally, yet you can likewise pay Rupees to step up any soldier you've opened at the Military Training Camp (whenever you've opened that administration). The going rate is shockingly sensible and there's no advantage to pounding an under-leveled character 'physically' through doing combating, so utilize the Military Training Camp — that is what it's there for.

Consider your catch pounding

It's simple and pleasant for newcomers to just crush the 'Y' and 'X' fastens haphazard, yet you'll get substantially more out of the battle in the event that you focus on the combos. They're not confounded; hit 'X' after a progression of standard hits and you'll release a solid assault that shifts relying upon the quantity of catch squeezes that preceded.

Once more, you can openly test in the Meditative Training territory so go there to try out moves as and when you open them and discover which combos you like best. It's not just about hacking-and-slicing, you know — it's tied in with hacking-and-cutting in style!

Utilize the Sheikah Slate rune capacities

The Sheikah Slate is an amazing little gadget. Holding down

'R' will slow time and let you pick between the four rune capacities:

Magnesis ('B') — utilize an attractive capacity to toss metal trash and divert foe weapons back on them

Cryosis ('A') — make a square of ice to help you in battle

Far off Bombs ('Y') — whip out some shining blue bombs to separate adversary safeguards

Balance ('X') — 'freeze' an adversary to a spot and energize a huge active assault through numerous strikes

You can't simply spam these assaults on account of a short cooldown meter that shows up between utilizes, however they're amazing assets in your munititions stockpile. While foe weaknesses to these rune assaults are motioned with symbols, recall that you can utilize them whenever — they're an extraordinary method to work on a baddie's wellbeing bar whenever, so remember about them.

Utilize your Special Attack deliberately

Watch out for your Special Attack meter(s) in the upper left corner of the screen. Whenever you've squandered enough snort foes it'll enlighten with 'A' in the center and squeezing the relating catch will start a monstrous screeb-clearing Special Attack.

Prudent utilization of this can make short work of even the hardest adversaries, and with a little technique you can utilize it to evade fatal assaults, as well. Actuate a Special similarly as a Guardian fires its laser at you and you'll accept no harm as the breeze up assault activity plays out.

Remember that you can repel

Recollect that Link and certain different characters can repel, and it very well might be much snappier or simpler to repel a strike and open up an adversary's frail point measure than to evade an assault (which opens up the Flurry Attack alternative) or sit tight for a flagged opening for a Rune assault.

How would you repel in Age of Calamity? Hold down 'ZL' to shield and hit 'Y' as an adversary's assault is going to hit. The adversary will be paralyzed and their feeble point measure uncovered. Stall out in.

Break each container and box...

Around each level you'll see fragile boxes and barrels — it merits going a tad out of your approach to break a couple, particularly in the event that you appreciate intertwining weapons. The Blacksmith is glad to take on whatever venture you like, however he'll need coin for his administrations. All things considered, he's not intriguing in coin — he just acknowledges rupees, and the money found in containers will come in helpful, particularly toward the beginning of the game.

You may likewise discover a Korok covering up in a case, as well.

Despite the fact that the catch brief to open chests will show up, you can securely overlook it and all the treats held inside will mysteriously get away and go into your pocket. This goes the equivalent for the rupees in boxes and any things dropped by adversaries or through natural demolition. Each one of those things will naturally home in and pursue you any place you go, so don't sit around rushing to gather them.

A few chests — ones explicitly concealed that you'll discover now and again are the exemption for this standard and should be opened physically with 'Y'. Something else, let the riches come to you.

Circuit weapons to make them more grounded

Weapons, similar to characters, can be stepped up. Just head to the Blacksmith's and he'll offer to combine any of your weapons together to settle on your base decision more grounded.

You can likewise coordinate weapons with similar advantages (demonstrated by uncommon symbols or 'Seals') and improve your top picks with exceptional buffs.

Incidentally added a Seal you don't care for, or need to supplant it with an improved rendition? Don't worry about it — later in the game you'll open the capacity to eliminate all the Seals (or chose ones), so you'll have the option to mold the ideal edge, club, skewer or other cutting weapon. Blast.

Recollect your basic bars

On the off chance that you can see Wizzrobes on the guide, it's perpetually worth chasing them down and crushing them to top off your natural bar meters. A speedy impact from your Fire Rod can make short work of ice adversaries, and it very well may be an extraordinary method of bringing down bigger foes right away.

Cook suppers for pre-fight buffs in case you're experiencing difficulty

By finishing Quests you'll open plans that can be cooked before a fight and which offer rate buffs on development speed, assault strength, cooldown speeds and numerous different factors. On typical trouble you may not need them, particularly prior on, however in a portion of the later difficulties and challenges you'll require each edge in fight you can get.

On the other hand, thump the trouble setting down

There's no disgrace in thumping the trouble level down in case you're experiencing genuine difficulty. To do this, head to the Options part of the title menu and change to a simpler setting. You can likewise check the trouble level you've finished an assignment on by checking the Battle Info on the principle map screen (hit 'X' whenever you've chosen a Subchapter or Challenge to get to it).

Utilize the 'L' and 'R' catches to get to Services, Chapters, Challenges and Quests on the overworld map screen

From the outset, you'll experience no difficulty seeing glimmering symbols on the overworld guide of Hyrule, however after a short time you'll have a larger number of symbols littering the guide than you'll have the option to stay aware of.

Luckily, hitting the guard catches ('L' and 'R') cycles between menus empowering you to get to Services, Chapters, Challenges, and Quests significantly more without any problem. These are conveniently partitioned by means of Recommended Level and Character, and they likewise show as 'Finished' whenever you've ticked them off. Along these lines, try not to strain your visual perception searching for a minuscule image on the guide — hit those guards and get where you have to go quick.

Utilize your Sheikah Sensor to discover the materials you need

You'll before long open the capacity to find ruins required for Quests by utilizing the Sheikah Sensor. By hitting 'X' over a Quest symbol you wish to finish, areas and Challenges that give the vital materials are featured in green. This implies you don't have to look to discover the prizes you need individually. You can even open the capacity to utilize the sensor on in excess of two objective Quests. You'll likewise get a warning mid-fight when you've gathered the imperative materials. Convenient!

The Sheikah Sensor has its constraints, however. Certain basic materials might be found in a wide range of areas, and actuating the Sheikah Sensor may bring about many symbols 'pinging' green, particularly in case you're utilizing it for at least two Quests. All things considered; it tends to be extremely valuable for finding an uncommon material.

Recollect you can purchase materials from Stables and different shippers, as well

It's additionally worth recollecting that you can frequently purchase the materials you need from providers over the overworld map instead of head into fight for them. This is maybe generally obvious with food things — Tabantha Wheat, Hylian Rice, Fresh Milk, Fish, and so on however you'll additionally discover Keese Wings, Octo Balloons, valuable jewels and substantially more available to be purchased.

Battling is the essential method to develop your load of these materials, however recollect that you can buy a large number of them with rupees and it's frequently a lot faster to purchase a couple that you're missing than head into an extensive fight just to snatch some Hyrule Herb.

The Sheikah Sensor will feature any opened dealer on the guide conveying materials you need, so put that to utilize.

Switch among characters and spot them deliberately.

Almost immediately in Hyrule Warriors: Age of Calamity, you get the choice to switch between playable characters on the fly. While it's incredible enjoyable to evaluate each new character in fight, you additionally can call up the guide and request your characters to various pieces of the combat zone. Since you can switch between your characters whenever, this implies that you can deliberately put them in various corners of the guide and speed between them to accomplish area explicit destinations quicker. In spite of the fact that, for harder battles, you might need to get back to the group together.

Watch out for Korok seeds.

Much the same as in Breath of the Wild, dubious articles or areas in Age of Calamity can contain the annoying Koroks! It could be more hard to monitor the little green animals while you fight away, however in the event that you set aside the effort to investigate and watch out, you'll make certain to locate some in cases or on scaffolds, for example. Gathering Korok seeds will again permit you to extend your weapon stock spaces, making it certainly justified regardless of your chance to look around each alcove and crevice of the guide!

Connection crush (for things)!

Some old propensities never kick the bucket, and this remains constant for Age of Calamity. Breaking different things thronw over the guide, for example, cases, can normally yield some helpful recuperating things and elements for making. The cycle is much more straightforward in Age of Calamity, as any dropped things essentially fly directly to you. Make certain to release Link's internal danger as snatching these things will give you an abundance of rewards like wellbeing things to use during battle and the capacity to make nourishment for assault and wellbeing buffs. Even better, any rupees you find can be utilized to gain new weapons, weapon overhauls, or even to pay to step up your characters and evade pointless pounding.

Try not to spam assaults.

It pays to consider your moves in fight. Characters like Impa can go through foes to manufacture her number of useable clones. Uncommon rune capacities like Remote Bombs and Cryonis can be utilized to disperse gigantic groups. Avoiding small supervisor assaults finally opens up the opportunity to utilize a moderate movement whirlwind assault (similar to in Breath of the Wild) that can break the adversary's safeguards. Moreover, smaller than usual managers will frequently have transmitted images to show their shortcomings directly as they end up for a major assault, so exploit those minutes. At last, each character additionally has exceptional combos that would all be able to be helpful in their own right, for example, utilizing X and afterward Y to make Link shuffle his adversaries in air.

Brain your environmental factors.

As referenced before, there are a lot of boxes to break for rupees and fixings in Age of Calamity. Notwithstanding, there's additionally significantly more to every region of

Hyrule than this. Trees can be brought down to get wood, and chests can be found containing weapons as well. Foe stations that aren't important for the fundamental goal can be found off in an unexpected direction and handled to acquire extra XP and plunder.

Focusing on your current circumstance helps in battle as well. Utilizing Cryonis in a waterway will freeze any close by foes, Magnesis can transform metal boxes into weapons, dangerous barrels will harm scaled down supervisors on the off chance that you can draw them sufficiently close, and you can commence any divider or vertical surface for a speedy escape by means of paraglider.

Do the sidequests.

When you access the Hyrule world guide, you'll rapidly see some little yellow symbols. A considerable lot of these are discretionary side exercises that let you exchange your procured things for an assortment of helpful advantages. A portion of these can give a character an additional heart or an all-encompassing combo assault, while others can give you new food plans for detail buffs and even admittance to shops like the smithy. Make certain to focus on what is accessible on the guide after every mission so you don't pass up a large group of extraordinary updates.

Symbols for journeys you can finish right presently will

shine. On the off chance that you don't have the materials to finish a sidequest yet, utilize the Shiekah Radar opened from the Needed: Researchers! journey right off the bat in Chapter 2 to be told which stage(s) will allow you the materials you actually need.

Finishing sidequests is particularly significant right off the bat when the suggested level for missions bounces from around 5-6 right to 16! Take advantage of what's accessible to you in the early game and give your gathering of characters the most ideal stuff for the battles to come.

Take advantage of those menus.

Between missions, you can alter your characters. Changing your weapons and burning-through nourishment for lifts to wellbeing and assault power are each of a crucial piece of how the mission may play out. A speed-boosting dinner could be exactly what you have to get away from the impact of a Guardian (in case you're bad at repelling), or you probably won't understand you had a more grounded weapon accessible to you.

You ought to likewise be aware of the gathering you bring into missions. View their combo assaults and special moves, and even check what materials will be reachable from the mission ahead. Information is force, and becoming acquainted with the interesting qualities of each character and what things you could pick up to later use in a sidequest you've had your eye on is fundamental.

Hyrule Warriors: Age of Calamity - Launch Trailer

Hyrule Warriors: Age of Calamity delivered yesterday only on Nintendo Switch and our Gamereactor Warriors Stefan Briesenick and David Caballero have been playing endless hours as of now to give you, other than the game's audit, a progression of helpful hints and deceives.

This is a musou game, valid, yet not your run of the mill as it's likewise been intensely altered dependent on The Legend of Zelda: Breath of the Wild's reality and frameworks. As such it offers players a lot of cutting-edge procedures alongside some extremely remarkable moves and apparatuses, including Sheikah Rune powers and rudimentary activities. Here are a few pointers, insider facts and proposals.

Breaker Weapons, produce better battle instruments

In Hyrule Warriors: Age of Calamity, there are basically two different ways to get more grounded: By either picking up experience focuses or by manufacturing better weapons. When you have opened the Hylian Blacksmith Guild, you can soften down undesirable weapons there to fortify your arsenal. Other than higher harm esteems, this cycle opens different detached extra points of interest (as Seals) and in the event that you combine similar impacts with one another, you even increment the comparing details.

Since some legends can prepare a few parts of weapon (which, coincidentally, influences the character's playstyle just as their combos), you should try to consistently combine a similar weapon type together. For instance, fundamental saint Link can prepare a short blade, a two-gave weapon, or a lance. On the off chance that you liquefy blades into his lance, you will gain far less ground, your rewards will be brought down and the entire cycle will get truly costly.

Hyrule Warriors: Age of Calamity - Chapters 1 and 2 Champions Gameplay

Softening modest weapons into your powerful Master Sword doesn't acquire a lot to the table the since quite a while ago run, since you need excellent/level hardware for some genuine advancement. Along these lines, either intertwine awful material or sell it straightforwardly - there will consistently be more. Later in the game, you will open more alternatives at the metal forger, such as restoring rusted weapons with Octo-Polish or changing undesirable uninvolved impacts. This normally comes at the cost of a ton of rupees and it isn't generally worth until you arrive at the later periods of Age of Calamity, where you need to additionally practice your #1 characters much more.

You ought to become acquainted with the produce as ahead of schedule as could be expected under the circumstances, as every one of your principle warriors needs a skilled weapon. Be that as it may, overhauling is probably going to tear an opening in your funds, so perhaps center around just a couple of characters at that point, rather than having everybody modern. Before you toss all that, you have into the smithy, consistently check your pockets cautiously. A few weapons have a high exchange worth and you ought to consider that alternative to top off your rupee pack.

On the off chance that you head out with a similar character consistently, her or his weapon sack will top off decently fast. Hestu can broaden your pockets, yet to do this you need Korog seeds, which arc all around shrouded away in the mission missions. Later in the game you can basically observe where to check for these collectibles, yet you'll have been playing for some time by at that point. In the event that you played The Legend of Zelda: Breath of the Wild, you can probably figure a portion of the spots you need to search for these small animals, so keep your eyes open.

Avoid and square

A few adversaries can require a long time to beat, especially when you're not sufficient yet. In these cases, and all in all, you better time your avoids (B) and squares (ZL) right. Both will open adversaries' protections for longer combos, and after some preparation you'll turn into a quicker, more solid fighter who doesn't rely a lot upon extraordinary forces.

Use Sheikah Runes against extraordinary assaults

The Sheikah Rune powers work with natural components and in battle, however as opposed to utilizing them all the time unwittingly, whenever they're reloaded you better use them deliberately, particularly with foes greater than your regular snorts.

At the point when one of these bigger adversaries begins a more explained or unique assault succession, a roulette will incite on screen rapidly to stop on one Sheikah Rune (a lock, a magnet, a snowflake, a bomb), and that is the Rune you need to use to counter them. In case you're effective (in addition to the fact that you have to be snappy and learn designs - you additionally need those Runes to be chilled off and prepared to utilize once more), you will both stop their assault and break their protection meter faster, which is an unquestionable requirement to complete foes, particularly in case you're in a rush. Accordingly, this halting/countering procedure is one of the absolute first you have to dominate, as you may not generally care from the start. Thusly, overhauling your Sheikah Runes must go high on your need rundown of Quests.

Rudimentary, my dear hero

In the event that Runes are utilized with R, at that point fire, ice, and power rudimentary bars enter a stone paper-scissors kind of game when in battle with L. You have to top off your poles by executing rudimentary adversaries (the Wizzrobes are the most self-evident, however different foes, for example, Moblins and Lizalfos are fueled/altered by the components) and by finding rudimentary shards.

Same likewise with Runes, don't squander this irrationally. We counsel you spare a few components on the off chance that you need to confront a contradicting component foe later on a given mission. Fire beats ice, ice beats fire... what's more, lightning is super helpful to stunt adversaries, way more so on the off chance that they're near water or metal. Truth be told, you should glance around and check whether the ground can be singed, for instance, for better impact. Another tip is that you spare some Rod power as an extra asset against the most safe and irritating managers.

Incidentally, adversary types likewise contrast incredibly from each other. Lynels, for instance, shoot electric bolts in any structure, yet only one out of every odd subspecies spill fire balls at you or run you over with a lance. Shiekah capacities are generally combined to these practices, so they will likewise vary. Once more, be careful and figure out how to respond to these examples. At that point you can do the fitting counter-responses quicker and break openings in the foes guards.

Effective Economy: Invest your rupees carefully

In the event that your wallet is coming up short and you're lacking in rupees, it very well may be beneficial to sell a portion of the more generally assembled beast materials you find all through the missions. Nonetheless, note that you may wind up deprived for explicit material later in the game and accordingly, having a base gracefully of everything is a smart thought. If all else fails, don't be reluctant to replay a generally finished mission, since this encourages you gain experience focuses (away to not spend a rupee step up characters at the Military Training Camp) and it likewise places some more assets into your pockets. Tirelessness doesn't do any harm, on the grounds that Hyrule Warriors: Age of Calamity is a broad game that will keep you occupied for some time in any case.

It additionally bodes well to work a little towards the new next story mission you are going to handle, as they ordinarily set aside a more extended effort to finish (half or even an entire hour rather than your run of the mill 10 brief Challenges). Attempt to change the degree of your soldiers before you head there - the Military Training Camp will help on the off chance that you have the vital measure of rupees, yet there's consistently a level cap. Likewise, cook something scrumptious with the goal that you can get significant extra points of interest like extra XP, more front line explicit crude materials and the preferences. On the off chance that you ranch for an explicitly crude material, change your feast and, if vital, your gear, as well. What's more, consistently recollect: Large hordes of adversaries are not just useful for your experience bar and to rapidly top off your extraordinary assault, however they assist you with getting important assets that may prove to be useful later.

Records make Hyrule Map function as a helpful menu

As the game advances, the world guide turns out to be exceptionally packed, making route more troublesome. With L you go to the "Administrations" tab, which records the main focal points just as shops. Experience this rundown routinely to recharge uncommon assets (additionally, the vendors renew their stocks after each fruitful mission).

In like manner, with R you access Chapters (story missions), Challenges (side missions arranged by suggested levels and furthermore Divine Beast Battles, and Quests (tasks you ordinarily pay with assets, and here you discover some as significant as overhauling Sheikah Runes, improving partners, or opening and upgrading Services).

Keep in mind: in case you're searching for that tricky material, simply allocate it to the Sensor in this menu, and the connected fights will be featured on the guide. Isn't it cool?

It's a mystery to everyone

Some more drawn out story missions permit you to go Onward with a few heroes, however they may likewise incorporate more than one chief or require some particular capacities. A few times it's insufficient to meet the suggested level, so you should test the ground and afterward return to pound/level up or to pick another saint. This applies to explicit difficulties too, for instance when you can't be hit or when there are time obliges. Additionally, for certain supervisors you enter alone with one playable character, despite the fact that you began the mission with additional. A master tip here is that you gather all the components shards and all the nourishment for that fundamental character who will take the enormous battle.

Another helpful, efficient stunt is that, have you seen those wooden watch towers? They typically have a few foes and an uncommon skull chest with an uncommon thing, correct? All things considered, the quickest method to clean it is with a bomb starting from the earliest stage: be in a split second annihilated, the fortune prepared for you to gather.

Other than this, and same similarly as with Breath of the Wild, continue perusing the tips appearing during stacking screens, as you'll find out about some truly cool stuff. What they won't let you know is that, on those screens, you can utilize L/R, B and X to control the Small Guardian. It's so charming!

The 5 Best Things About Hyrule Warriors: Age Of Calamity

Best: Tools

The instruments in Breath of the Wild worked uniquely in contrast to any game in the arrangement before it. While it was cool to utilize them to break the game with unusual material science and they could be used in battles, they weren't explicitly intended to be hostile.

Not exclusively are the material science generally flawless in Hyrule Warriors: Age of Calamity, however apparatuses like bombs additionally sneak up suddenly. Each legend likewise works these devices in an unexpected way, so no one can tell how things will turn out until they are attempted.

Most exceedingly awful: Not As Robust As Hyrule Warriors

As incredible as the devices may be, there is an absence of substance in Hyrule Warriors: Age of Calamity. There aren't the same number of saints to look over and it is more limited than its archetype, Hyrule Warriors, even with the discretionary missions.

The first arrival of Hyrule Warriors was stuffed with content gratitude to the Adventure Mode, and the Switch rendition dispatched with much more additional items. It's a disgrace to see the spin-off not opponent the Wii U unique.

Best: Traversal Options

There are a bigger number of approaches to go than just on foot. Pretty much every character can jump out a lightweight flyer. While it tends to be precarious to pull off, the alternative is there and very welcome. One exemption is Revali because he can fly, making the lightweight plane repetitive.

Connection can likewise slide on his shield while hitting X in his running activity. It's a little triumph for Hyrule Warriors: Age Of Calamity however the lightweight planes alone cause it to feel more like Breath of the Wild.

Most noticeably awful: Still Want Horses

Tradition Warriors and its numerous side projects ordinarily have something equal to a pony to get around. Ponies were deficient in the first Hyrule Warriors they're actually absent here.

It looks bad since The Legend of Zelda arrangement has consistently been acceptable about fusing ponies, including Breath of the Wild. Skimming around is fun yet nothing can match a pony regarding speed.

Best: Voices Are Back

Other than the Philips CD-I barbarities and the '80s animation, The Legend of Zelda arrangement has been quiet. That is up until Breath of the Wild. The voice cast returns for this prequel, which ends up being both a decent and terrible thing.

From one viewpoint, this is a positive advance as the principal Hyrule Warriors didn't have voices. Then again, a portion of the voices, as Daruk and Zelda, aren't unreasonably incredible. It's superior to nothing.

Most noticeably terrible: No Voices In Battle

Not everything in Hyrule Warriors: Age Of Calamity is voiced which incited this section to dock the game a couple of focuses. While in fight, text boxes will spring up. While now and then it's only there to add flavor to battle, different occasions they contain significant subtleties for a mission.

It's sort of difficult to peruse in the corner while killing through several soldiers. For what reason aren't these voiced like in the remainder of the game?

Best: Weapons Don't Break

As celebrated as Breath of the Wild seemed to be, there was one problem that was particularly polarizing with the two fans and pundits: the weapons breaking. Fortunately, you don't need to stress over that in Hyrule Warriors: Age of Calamity as it astutely disregarded this ongoing interaction component.

Truth be told, players can even circuit weapons together to level them up. It's not the most ideal approach to spend rupees but rather the choice is there.

Most noticeably terrible: No Costumes For Anyone Besides Link

Connection is the main character in Hyrule Warriors: Age Of Calamity that can prepare ensembles. These outfits are generally restorative and hold practically zero an incentive as far as boosting details.

For what reason didn't any other person will spruce up? So far as that is concerned, for what reason are the greater part of Link's decisions directly from Breath of the Wild? It fails to impress anyone that is without a doubt, however it is fairly justifiable. Nintendo probably wanted to keep up a particular search for a large portion of the characters, while actualizing some cosmetic customization through Link.

Best: The Map Layout

From the outset, the guide of Hyrule may look overpowering. With each mission complete, one, or twelve additional symbols will spring up.

For those that solitary consideration about the story, those symbols are gigantic so they are anything but difficult to track down. For those that need to investigate, at that point there are bunches of decisions. While the greater part of them reduce to give X measure of a thing to a resident or murder X measure of foes, the fact is exploring is straightforward.

Most exceedingly awful: Graphics Look Dated Already

This game is fundamentally a mod on top of Breath of the

Wild's guts and it shows. Having the same number of adversaries on screen without a moment's delay is truly pushing the Nintendo Switch's motor as far as possible.

Hyrule Warriors: Age of Calamity is emulating the 2017 game however not out and out creation it as lovely in view of the number of foes on screen. The stoppage additionally doesn't help. All things considered, it's prettier than most Dynasty Warriors games.

TIPS AND TRICKS

Get mipha first

In section two, you can start the way toward opening the four Champions in Hyrule. You can add Daruk, Mipha, Revali, and Urbosa to your gathering in any request, however you ought to organize making a beeline for Zora's Domain first to get Mipha.

She's a significant individual from your gathering, particularly in case you're playing on harder troubles, since she's the main character who can recuperate others. While you can discover food over each guide to top off your wellbeing whenever, it's a restricted asset. You scarcely run over it in higher troubles, so Mipha's your smartest option to remain alive on the off chance that you plan on playing the game on an all the more testing setting.

When battling close by her, the pools of water she makes with her assaults recuperates partners who are remaining in them. Carry her to any supervisor battles alongside different characters to build your harm yield and inactively recuperate different individuals from your gathering.

At the point when her exceptional assault is completely energized, she lets out a gigantic influx of water around her that harms any foes in the sprinkle zone. This move likewise tops off a huge lump of her wellbeing and the equivalent goes for party individuals trapped in waves.

Do quests to get longer combos

The further you get into the game, the more character-related missions you'll open. These simple to-finish missions have symbols dependent on each character in your gathering. To finish them, you should turn in various arrangements of things dependent on each journey. These are typically things that you acquire in the wake of beating foes, finishing missions, or buying them from shops.

The sorts of missions you can finish will either give your characters more wellbeing, or they'll add more assaults to their combos. You can sort out what you'll get by checking the prize recorded at the lower part of the journey depiction. Since the greater part of Hyrule Warriors: Age of Calamity expects you to knockout huge number of adversaries, having longer and more grounded combos makes this cycle simpler.

At whatever point new character-explicit missions show up on the guide, total it as quickly as time permits. In the event that you can finish a mission with the materials you presently have, the symbol will be bigger and throbbing. In the event that you don't have the materials to finish it, utilize your Sheikah Sensor to discover the things you need.

Do combat challenges as link

Battle difficulties are an oftentimes opened kind of mission, signified by their blade molded guide symbol.

The reason for these more modest fights is to acquaint you with certain battle mechanics, for example, adversaries with various natural sorts. Finishing them will assist you with seeing more about battle, however you'll likewise wind up giving the character who completes them more experience.

For the vast majority of these battle difficulties, you can do them as any character you like, yet do them as Link. He's is needed in most fundamental missions, which implies he'll probably have the most experience out of any of your gathering individuals at some random time.

Do this, and Link may have a lot more elevated level than every other person. This is something worth being thankful for. At the point when you level up characters by utilizing the Military Training Camp, you can just expand their level to that of the most noteworthy part. In the event that you guarantee that Link consistently has a lot more elevated level than every other person, at that point you entire gathering benefits (as long as you have the rupees to spend at the instructional course).

Gain proficiency with the sheikah rune differences between characters

Each character in the game can utilize the four Sheikah Rune assaults: Stasis, Remote Bomb, Cryonis, and Magnesis. For every colleague, and Magnesis act the very same when enacted. Be that as it may, Remote Bomb and Cryonis work distinctively between each gathering part.

At whatever point you get another character, it merits taking a gander at their move rundown, and afterward squeezing directly on your thumbstick to see the clarifications for their diverse Sheikah Rune capacities. No two are indistinguishable, so it helps realizing how differed they are.

Everybody has an alternate Remote Bomb and Cryonis assault Hyrule Warriors: Age of Calamity

For example, when Link tosses his Remote Bombs, you'll gain the capacity to point while he throws out a few preceding hurling a huge one to wrap up. Then again, Zelda's Remote Bomb turns into a controllable robot that latently tosses little bombs around it prior to detonating.

It's essential to know the distinction between these assaults since certain adversaries like managers and Wizzrobes will be dazed when hit with a Remote Bomb or Cryonis. Realizing how these two explicit assaults work for each character will assist you with being more compelling in fight when it comes time to utilize them to land a conclusive hit on a major foe.

Tenderfoot Battle Tips

You don't need to catch each station. Doing so is extraordinary for gathering more materials, developing your unique assault, and shielding foes from pursuing you around the guide. All things considered, most are not needed to advance the current mission.

Feeble point assaults are commonly the most remarkable.

You can perform one when you totally pulverize an enormous adversary's red powerless point measure. These are normally more impressive than your unique assault, particularly late.

Spare your Sheikah Rune assaults for counters. At the point when you're battling a named adversary that you can bolt onto, sit tight for the image of a Sheikah Rune to show up over their head, at that point utilize a similar assault to shock them and power the feeble guide check toward show up.

Request your characters to various areas on the guide! In the event that there are numerous waypoints on the guide, open up the menu and utilize the guide to drag any extra playable character to any area. This will let you keep doing whatever you might feeling like doing while the other character runs off to the point you chose. Whenever you're finished with whatever target you were finishing, you can switch straightforwardly to the next character.

Protecting is acceptable, however have you attempted to avoid all things being equal? In the event that you evade at the perfect second, you'll enact a whirlwind assault where the adversary eases back down and you accelerate to convey a few snappy assaults in succession. This likewise uncovered the powerless point check!

In case you're similar to a large number of us, you're likely sparing your essential poles for the perfect second. Try not! Wizzrobes and other natural based foes give ammunition to your fire, ice, and electric bars and these adversaries show up in essentially every level. Try not to store them!

Materials, rupees, and chests will fly out after you rout a foe or cut up some grass in the climate. Try not to stress over gathering them! They'll twist to you in almost no time on the off chance that you don't contact them first.

Before a Battle Tips

Each playable character gets a preparation challenge soon after being opened. Utilize those genuinely straightforward difficulties to check new characters out. Take a gander at the combo screen and attempt each and every combo. These combos for the most part comprise of hitting Y a few times and adding X to the furthest limit of a line of Y combos to play out a solid assault. Every one accomplishes something else!

Need to support the EXP of a character? Take a stab at concocting a dish that supports the EXP rate got in a level. There are additionally plans for expanded harm and quicker running velocity.

Turn on adversary wellbeing bars for all the more modest adversaries that aren't named and can't be bolted on to. Simply bounce into the choices menu and switch the slider.

Complete journeys for extra rewards and new administrations! These will let you buy more materials, meld weapons, diminish rune revive time, and bounty more.

At the point when you combine weapons, watch out for the seal of the weapon. This will concede that weapon a unique trait like an expansion in beast material droppings or a higher harm yield to adversaries noticeable all around.

Tips for Finding Materials

Cut grass and other vegetation in levels! These will drop materials sometimes.

Wooden cartons and different articles can be annihilated. These will regularly drop rupees.

Prior to heading into a skirmish of challenge, take a gander at the fight subtleties screen to see a rundown of level-explicit materials that show up more frequently.

In the event that you opened any shops through finishing journeys, observe that they reset their stock each time you complete a mission or challenge. The materials here don't change.

On the off chance that you opened any pens or shippers by finishing missions, observe that only one out of every odd trader will show up without fail. Play a mission or challenge to basically reroll your odds of a vendor showing up.

Have you seen that gleaming hare staying nearby segregated zones in specific levels. Moving toward it will make it hurry off, yet on the off chance that you hit it from a good ways (with Link's bolts, for instance) at that point rupees adding up to 100 will blast out.

Get your hands on the Sheikah Sensor! Right off the bat in Chapter 2 you'll open the Needed: Researchers! mission. Complete it and you'll pick up the capacity to enlist up to two journeys where you're feeling the loss of a necessary material. This will cause levels containing said material to throb a green ring. You'll be told in the wake of finishing a level in the event that you have enough of the material to finish the mission! Another journey in Chapter 4 permits you to enroll up to three missions.

There are cooking plans and weapon seals that help the drop pace of materials. Post for these and use them in missions!

Know, essential chests in missions will return while the fancier chests with uncommon materials won't.

Later on in Chapter 5, journeys will be opened that once finished will diminish the cost of things in shops.

Korok Searching Tips

Koroks can show up as unmistakable yellow blossoms or a pinwheel. They likewise stowed away in wooden boxes and some even take cover behind broke rocks that can be devastated with Remote Bombs. Likewise watch out for pivoting Korok inflatables!

Arrive at Chapter 4 and you'll open the A Most Peculiar Korok journey. Complete it and the quantity of Koroks found in each level will show up in the fight subtleties screen.

In the event that you need more Korok Seeds for a mission, don't worry! A few difficulties and missions will open extra Korok Seeds despite the fact that this doesn't occur until some other time in the game.

Progressed Tips

Get noticeable all around! You can keep on performing ordinary assaults noticeable all around and you can even end with a solid assault by hitting X and hammering once again into the ground.

You can get noticeable all around by hopping off fundamentally anything. Run into a divider, a stone, a barrel, or whatever you can't break and hit B twice to get into the air and begin floating. Some solid assaults and Sheikah Rune assaults will help you up as well.

Attempt to hold X when playing out specific combos. For certain characters, particularly for Link, hold X as you play out your solid assault will permit you to play out a second and exceptional solid assault once you let go.

In the event that an adversary has their frail point measure uncovered, consider actuating Stasis to freeze them and give you a couple of additional seconds to annihilate the feeble point check. Nonetheless, a few characters have a programmed follow up move for Stasis so it's not as powerful with all characters.

Utilizing an essential bar on an adversary won't just paralyze it yet uncover and debilitate their powerless point check. You can do this whenever, regardless of whether the foe just shrouded their powerless point measure. This is extraordinary for supervisors or huge foes like Lynels.

Figure out how to repel a Gaurdian's laser! As the Guardian locks onto you with its red look, hold your shield button (L) and sit tight for it to strike. Similarly, as the laser is going to hit you, hit Y to mirror the laser directly back at the enormous monster.

Don't Just Use Link

As enticing as it very well may be, don't simply utilize Link. He's probably the best character in the game, if not the best all things considered, yet Age of Calamity anticipates that players should utilize pretty much everybody. Different endgame missions will require either a full gathering, or characters who aren't Link, conceivably street obstructing the individuals who have been zeroing in on the Hero of the Wild.

Of the primary playable characters, just Hetsu, the Great Fairy, and Monk Maz Koshia can be serenely disregarded. The rest should all be leveled. Connection, Zelda, and Impa are the smartest options for overcoming the fundamental story, while each Champion should preferably be in their 60s by their last clump of side missions.

Wire Weapons

Depending on base weapon drops in Age of Calamity is a fiasco already in the works. While it is conceivable to get some solid Lv. 1 weapons, the reality they won't have any seals (apparently the game's ability framework) implies they won't be half as valuable as they should be. All the more significantly, their assault detail will be extremely low come mid-game.

It's significant that players update their weapons not exclusively to expand base assault, yet to add abilities that work off characters' play styles. While weapons can be at first leveled up to 20, this can be pushed to 25 and at last 30 throughout the span of the story. Raising a weapon's maximum level additionally nets them new seals.

Restock On Materials Regularly

Age of Calamity's journey framework is genuinely clear, essentially expecting players to convey materials at whatever point they have enough. While playing through the game normally should develop a good load of material, the numerous corrals littering Hyrule will consistently be the smartest choice for restocking materials.

RELATED:

10 Small Details In The Story Of Hyrule Warriors: Age Of Calamity You Missed

In the event that players have Rupees to save, it's never an ill-conceived notion to purchase more materials. Not exclusively will this moderate crushing (which will undoubtedly happen come mid-game,) merchants additionally sell fixings which can be utilized in cooking – apparently more basic in AoC than it was in Breath of the Wild.

Cook Accordingly

On that note, it's consistently a savvy thought to cook before a fight. Where cooking in Breath of the Wild must be done at flames, players pick what feast they will eat preceding a mission in Age of Calamity. Every dinner has its own buff which goes on for the whole mission, with players opening more supper spaces as they play.

All things considered, don't simply eat a feast to have eaten. Spare elements for stages where they'll truly sparkle. Getting an uncommon thing drop buff doesn't mean a lot of when a phase just has Moblins. Essentially, it's not savvy to eat a Rupee buff supper for the Coliseum stages.

Try not to Button Mash

It'll be enticing and even somewhat fun, however it's not insightful to fasten crush through Age of Calamity. While traversing the principle story won't be incomprehensible for any catch mashers, the side missions will end up being a major issue. Lynels are persistent, and Hinox appear to be intended to menace any individual who keeps button pounding.

Learn combos, set aside the effort to dominate a character' range of abilities and figure out their planning. Essentially beating the Y and X catches will bring about an ongoing interaction circle that never arrives at its maximum capacity. Age of Calamity's battle is truly loaded with profundity, however it won't be found by squashing catches.

Chain Specials Into Combos

Each character has their own Special that is set off with the A catch when their Special check is filled. An all around coordinated Special can be the contrast among life and passing on higher trouble modes, amazing smaller than expected supervisors or simply freeing waves from adversaries in a second's glimmer. Rather than simply jump starting out a Special once the check is full, change it into full combos. Not exclusively will the Special land harder, the lead-in can undoubtedly chip through a manager's shield, taking into consideration considerably more prominent harm gains.

As a matter of fact Stop To Give Orders

Dissimilar to the first Hyrule Warriors (and most Dynasty Warriors games,) Age of Calamity doesn't utilize unit the board. Or maybe, the emphasis is fundamentally on transforming Breath of the Wild into an activity game where keen players can overcome most any stage with a solitary character – yet few out of every odd.

Come Chapter 6, the primary story will begin tossing multitudes of foes at the player that must be dependably countered by requesting units to safeguard fortifications. This goes twofold for side missions, where a few late game missions test how well players can deal with a combat zone.

Investigate (When It's Safe)

Period of Calamity might be an activity game at its center, yet it includes a considerable amount of investigation. There's nothing on similar scale as Breath of the Wild, yet all significant guides include Koroks, Treasure Chests, and Easter eggs to discover. It's essential to consistently investigate another stage, seeing all that there is to offer.

Obviously, it's imperative to do so just when it's protected. Not exclusively complete a few missions include time limits, investigating when fortifications are enduring an onslaught is essentially requesting a game over. Make a point to investigate everything, except simply subsequent to getting out significant foes and arriving on a non-squeezing objective.

Do Side Missions As They Come Up

Don't simply keep up with the primary account. Period of Calamity is anything but an especially troublesome game, however the leveling adjusting will go of whack should players solely play the primary story. Making the most out of the game requires playing through side ridiculously up, acquiring new materials and step up the gathering all the while.

Leaving too many side missions immaculate outcomes in a frail center gathering of characters who will undoubtedly battle during the last couple of parts. More awful, players will pretty much need to depend on the late-game characters to get them by except if they need to granulate all the side missions they've left fixed.

Use Rods

Poles are a hero's closest companion, which makes it all the additionally baffling that most players don't utilize them. Not exclusively do Rods in a flash raise a supervisor's shield, they can be utilized to bargain genuine harm against the right essentially adjusted adversaries. Fire, Ice, and Thunder Rods go far in dealing with some of Age of Calamity's hardest difficulties. A solitary Rod can be all a player requires to tear through a Guardian or make light of a Lynel's danger.

HOW TO UNLOCK CHARACTERS

In Age of Calamity's first part, you'll work through the story as it builds up its position in the timetable. The subsequent part centers around gathering together different saints to help you in your excursion. All the fundamental characters have required initial missions to finish before they go along with you.

Extraordinary characters can be opened in later sections. These partners require the consummation of a few discretionary journeys and missions before they join your gathering.

How to open all characters

Hyrule Warriors: Age of Calamity includes a few playable characters from the Legend of Zelda: Breath of the Wild storyline. While you gain admittance to a portion of the cast directly from the beginning, opening the rest requires some work.

There are two arrangements of characters you'll open in the game: The four Champions from Breath of the Wild and a few exceptional characters from a similar title. The Champions and one of the unique characters become accessible as you play through the fundamental story, while the other party individuals include a few stages before they join your group.

In this Hyrule Warriors: Age of Calamity manage; we'll tell you the best way to open each character and what to do after you have them in your gathering.

CONNECTION/THE BATTLE OF HYRULE FIELD

Obviously, Link is accessible right from the beginning of the game. Truth be told, you'll be playing as Link in pretty much every significant story mission. In contrast to different characters, Link can utilize numerous weapons which changes up his battle style. He can utilize a blade and shield combo, two-gave weapons, and lances.

IMPA/THE BATTLE OF HYRULE FIELD

This mission likewise lets you play as Impa, the ninja-like champion. This initial mission permits you to unreservedly trade among her and Link. In future levels, you'll have the option to trade between more characters, yet in this stage you just have these two.

ZELDA/GUARDIAN AMOK

In the second significant story mission, you will control Zelda. This will support your gathering breaking point to three. Since this stage is bigger than the primary level, this will be the first occasion when you'll need to trade characters all the more regularly and issue orders, which is a basic hint we notice in our novice's guide.

THE CHAMPIONS: DARUK, MIPHA, REVALI, AND URBOSA

The four Champions (Daruk, Mipha, Revali, and Urbosa) are playable in Age of Calamity. To open them, you'll need to beat their particular missions in section two. Every one of these levels play out in an unexpected way, yet every one of them end with the capacity to play as each character once finished.

MIPHA/MIPHA, THE ZORA PRINCESS

To open Mipha, head east to Zora's Domain in Lanaryu. Her stage includes stumbling into her realm overcoming foes, having a manager battle with a Lynel, and controlling her Divine Beast, Vah Ruta.

Open Mipha first. Her solid assault and amazing unique assault move both have mending properties. Mipha's solid assault makes wellsprings on the combat zone. In the event that any of her partners are remaining in the water, they'll get a modest quantity of wellbeing. At the point when she does her sweeping extraordinary assault, the encompassing flood of water harms foes and furthermore reestablishes a lot of wellbeing for Mipha and her partners.

URBOSA/URBOSA, THE GERUDO CHIEF

To open Urbosa, head southwest to Gerudo Town in Gerudo.

Urbosa's stage will at first make them fight other Gerudo heroes until you get together with her partially through. At that point you'll assume responsibility for Urbosa while fending off an attack from the Yiga Clan. The stage closes with a supervisor battle against Master Khoga, in which you should crush him with of runes.

Urbosa's lighting-based capacities permit her to overcome huge influxes of adversaries or arrangement a great deal of harm to a solitary objective. Notwithstanding, to utilize her solidarity, you need to physically energize her lighting powers utilizing your extraordinary activity button. Doing so leaves her helpless, yet the result can be monstrous.

DARUK/DARUK, THE GORON HERO

To open Daruk, head upper east to Death Mountain in Eldin. His stage will make them run all over a fountain of liquid magma while overcoming different adversaries and an enormous Igneo Talus. You'll likewise safeguard his home by controlling his Divine Beast, Vah Rudania.

Daruk's solid assaults can gather various types of magma over the war zone, which he would then be able to detonate freely utilizing his one-of-a-kind activity. A portion of his assaults likewise make a shield around him that avoids assaults.

REVALI/REVALI, THE RITO WARRIOR

To open Revali, head over to the Tabantha Frontier, in the northwest district of Hebra. During your trip here and there the blanketed town, you'll take on a few Rito champions with a supervisor battle against Revali himself toward the end. When you rout him, he will join your gathering?

Revali's interesting move lets him make to the sky, giving him a different arrangement of moves while airborne. He can remain over the ground however long you need or until you get hit. His bow assaults and combos are likewise incredible for dispatching enormous quantities of adversaries rapidly.

HESTU/FREEING KOROK FOREST

To open Hestu, finish the initial segment of the "Liberating Korok Forest" mission. Whenever you've made a way into the woodland, your gathering will meet Hestu. When your group has the occasion to separate, he'll join your gathering. Finish the mission, and Hetsu will join your group.

Hestu assaults with enormous maracas and his one-of-a-kind activity permits him to call upon different Koroks that battle close by him. They will either latently assault different adversaries or become a portion of Hestu's assaults to reinforce his harm yield.

PRIEST MAZ KOSHIA/TRIAL OF THE ANCIENTS

In section four, three missions open up that require you convey a small bunch of materials. In the wake of finishing the journey prerequisites for "The Trial of Monsters," "The Trial of Stones," and "The Trial of the Mighty Stone," you'll open a test called "Preliminary of the Ancients." The fight sets you in opposition to the old priest, Maz Koshia. The fight itself isn't excessively testing, yet you should beat your adversary inside 3 minutes and 30 seconds.

Maz Koshia's assaults include calling enormous weapons with their solid assaults. These weapons at that point detonate, conceding the priest a portion of their energy. Convert this capacity to pummel the combat zone with a huge territory assault.

THE GREAT FAIRIES/FAIRY FOUNTAIN: BATTLEFIELD

In section four, a mission will open up called "Offering Help on the Road," which costs 4,500 rupees to open and access the "Preliminary of the Great Fairies." Thankfully, when this journey is finished, you'll get twice your cash back.

From that point forward, you'll open three difficulties in which you should clear enormous front lines and gather rupees for Great Fairies. When you complete the mission prerequisites for "Pixie Fountain: Deep Snows," "Pixie Fountain: Grassy Plains," and "Pixie Fountain: Desert Sands," you should overcome a Great Fairy toward the finish of each.

Doing so will open one final mission, "Pixie Fountain: Battlefield." For this mission, bring your most grounded character, ensure you have a lot of Wizzrobe wands, and be on point with your evades to get surge assaults. You'll have rout different huge foes inside an exacting time limit. Toward the finish of this mission, you'll have to vanquish one final Great Fairy. Bring her down, and every one of the four Great Fairies will consolidate into one character who joins your gathering.

As a gathering, the Great Fairies switch between assaults. Every one of the four Great Fairies will play out an alternate assault that influences the whole combat zone around them. The main exemption is the purple-haired Mija, whose exceptional activity makes an impermanent boundary around her.

There are likewise some exceptional character who open close to the furthest limit of the game, which we'll put toward the finish of this guide.

WHAT TO DO AFTER YOU UNLOCK CHARACTERS

After you open another character, two arrangements of levels show up on your guide.

The first is a bunch of missions with a solitary blade symbol. These character-explicit mission show you how to play as your recently opened character. The fundamental advantage of these difficulties is having the option to take new colleagues to a little field where you can try different things with their moves.

A bunch of missions will likewise open up that include a symbol dependent on each character. These journeys either permit you to raise the measure of wellbeing for that character, or they open more combo alternatives. These missions frequently expect you to turn in specific materials. In the event that you are experiencing difficulty finding the correct materials, at that point utilize the Sheikah Sensor to discover what you're absent.

RIJU/AIR AND LIGHTNING

Before you can recover control of Vah Naboris in this mission, you need to enable Urbosa to crush Thunderblight Ganon. Supporting you in this battle is Riju, a descendent of Urbosa who heads out from the future to assist you with winning the battle.

Riju carries on the custom of her progenitor by utilizing power to help her in battle. She likewise battles with close by her Sand Seal, Patricia. Together they ride over the war zone crushing into foes and projecting electrical discharges any place they go.

SIDON/WATER AND FIRE

As you battle to recover control of Vah Ruta, Mipha will battle in her battle against Waterblight Ganon. Amazingly, a future variant of Sidon — a similar one you meet in Breath of the Wild will join the battle to support you and the Champion battle against the chief.

Like his sister, Sidon battles with a lance and water-based moves. His one of a kind activity permits him to intensify his solid assaults in the event that you follow the on-screen lines. Hitting the novel activity button at the opportune time can transform Sidon's ordinary solid assaults into screen-filling attacks.

TEBA/AIR AND LIGHTNING

While Revali attempts to without any help battle Windblight Ganon on top of Vah Medoh, his substitution from the future, Teba, joins the battle.

As an individual Rito champion, Teba battles with a bow and bolt. The greater part of their assaults center around using the weapon with their exceptional assault permitting them to energize an incredible bolt volley the more extended the catch is held down. Their combo finishers with solid assault center around various bolt shots that stretch over the war zone.

YUNOBO/WATER AND FIRE

With Daruk battling to put down Fireblight Ganon on Vah Rudania, the Goron's descendent from the future, Yunobo, goes from the future to spare him.

Yunobo can initiate a similar defensive shield as Daruk, anyway his usage is extraordinary. While it can at present assist him with diverting harm, the obstruction can likewise control up his solid assaults. In any case, each time the force is initiated, it will improve an alternate solid assault.

ZELDA (BOW OF LIGHT VERSION)/EACH STEP LIKE THUNDER

Close to the furthest limit of this mission, Zelda's forces at long last actuate permitting her to utilize her Bow of Light. With this weapon prepared, Zelda's assaults change and she plays like a totally unique character.

With the Bow of Light as her primary weapon, Zelda's principle moves become Trifoce-replenished mystical assaults. These capacities can make rings of light around foes, making them take more harm from Zelda. Her extraordinary activity permits her to switch into Luminescent mode, where she can utilize her exceptional measure to convey enormous energy assaults.

EXPERT KHOGA/EACH STEP LIKE THUNDER

Expert Khoga's solid assault combo enders add to a pressure measure. On the off chance that this meter tops off, he'll pitch a fit and fall over and you'll have to squirm your left thumbstick to deal with him once more. Be that as it may, on the off chance that you eat a banana (by hitting the solid assault button without anyone else) and can finish the long eating movement, his pressure will be changed over into an uncommon meter for a restricted measure of time. In the event that the meter tops off while still under the impacts of his banana, hitting the novel activity catch will release an assaulted called the Big Glowy Blast.

RULER RHOAM/GREAT PLATEAU

The ruler can switch between two structures by hitting the remarkable activity button. His lord pretense permits him to convey solid, yet moderate assaults and his recluse appearance changing it up with quicker strikes. On the off chance that you press the remarkable activity button toward the finish of a combo that closes with a solid assault, King Rhoam will switch appearances while conveying an extra solid assault.

How To Unlock The Secret Ending In Hyrule Warriors: Age Of Calamity

There are two goals players need to finish the open the mystery finishing of the game.

Complete the Main Campaign: This remembers finishing the entirety of the story missions for every one of the 7 parts of the game.

Open Terrako: This is accessible in the post-game substance. The player should finish a few elevated level difficulties to open them. The player will likewise need to gather 50 Terrako Components littered all through missions in the game.

Presently, the player needs to beat the game again and the player will be welcomed with the mystery finishing just as another adaptation of the credits. Hyrule Warriors: Age of

Calamity is an incredible difference in movement from the first arrangement that recounts the tale of a mainline Zelda title. While the plot in Breath of the Wild was negligible because of the weighty accentuation on investigation, Age of Calamity gets a move on by giving players the foundation on every one of the victors and numerous different characters from this time of Hyrule. The whole game can likewise be played with a companion, making it one of the better helpful encounters on the Nintendo Switch.

HINTS & TIPS

New champions join at the normal degree of your gathering. In the event that any of your characters are seriously falling behind, they will keep down your new characters.

Level up those strays in the Military Training Camp in case you're not too attached to them.

Traders turn their accessibility each time you complete a mission or side journey. Watch out for this in case you're searching for explicit materials you can purchase from merchants to finish a mission.

Most trees, barrels, and cases can be crushed in the event that they are straightforwardly on the war zone. Devastate them to get materials or cash.

You don't have to stop and plunder the chest and materials that drop when you execute a chief. They will naturally move to your stock.

Connection's shield will stop a ton of assaults, particularly tossed things, yet he will make a sound like he's been hit or for a few, will be ablaze in spite of not being harmed by it.

Connection is awesome against Hinox in the event that he has the blade and shield. Having the blade and shield implies he actually has the utilization of his bow, and on the off chance that you point the bow at the eye of the Hinox.

Made in the USA
Columbia, SC
11 October 2023

24273566R10046